BALANCING CHEMICAL EQUATIONS WORKSHEETS

Over 200 Reactions to Balance

Chemistry Essentials Practice Workbook with Answers

Chris McMullen, Ph.D.

Balancing Chemical Equations Worksheets

Over 200 Reactions to Balance

Chemistry Essentials Practice Workbook with Answers

Zishka Publishing

ISBN-13: 978-1-941691-07-6

ISBN-10: 1-941691-07-2

Education > Science > Chemistry > Workbooks

Contents

Introduction

The art of balancing chemical equations challenges students' problem-solving abilities. It's more than just applying mathematics: It combines trial and error with reasoning skills.

Practice can help students develop fluency with balancing chemical equations. This chemistry workbook provides ample practice:

- Step-by-step examples with explanations.
- Over 200 chemical reactions to balance.
- Problems grow progressively more challenging and involved.
- Answers to every problem in the back of the book.
- One chapter devoted to pre-balancing exercises.
- A concise review of pertinent concepts and ideas.

This chemistry workbook isn't just for students. Anyone who enjoys math or science puzzles may enjoy balancing chemical reactions in the spirit of solving puzzles.

1 How to Balance Chemical Reactions

Definitions

Let us begin by defining terms that are relevant to chemical reactions.

A **chemical change** occurs when a new substance is formed with a different composition.

An example of a chemical change is when carbon (C) and diatomic oxygen gas (O_2) get together to form a new gas called carbon dioxide (CO_2). At the molecular level, two oxygen atoms are bound to each carbon atom in carbon dioxide. The composition of carbon dioxide is thus different from that of separate carbon and oxygen atoms.

A **chemical reaction** occurs when two or more substances undergo mutual chemical changes.

An example of a chemical reaction is the synthesis of sodium chloride (NaCl) from sodium (Na) and diatomic chlorine gas (Cl_2). In this case, sodium, chlorine, and sodium chloride undergo mutual chemical changes. The separate Na atoms and Cl_2 molecules change chemical composition in forming NaCl molecules.

The **reactants** of a chemical reaction are the initial substances. They react together to produce new substances with different compositions.

The **products** of a chemical reaction are the final substances. They are the new substances that are produced by the reaction.

For example, in photosynthesis, plants combine water (H_2O) and carbon dioxide (CO_2) to form carbohydrates, such as glucose ($C_6H_{12}O_6$), and diatomic oxygen gas (O_2). Water and carbon dioxide are the reactants (these substances react together), while glucose and diatomic oxygen gas are the products (these substances are formed by the reaction).

A **chemical equation** represents a chemical reaction in symbolic form, with the reactants added together on the left-hand side, the products added together on the right-hand side, and a yield symbol (→) in between.

An example of a chemical equation is:

$$NaOH + HCl \rightarrow NaCl + H_2O$$

The previous chemical equation represents the following chemical reaction: Sodium hydroxide (NaOH) reacts together with hydrochloric acid (HCl) to yield sodium chloride (NaCl) and water (H_2O). The reactants are NaOH and HCl, while the products are NaCl and H_2O.

The **terms** of a chemical reaction are separated by plus (+) and yield (→) signs.

For example, consider the following chemical reaction:

$$FeS + 2\ HCl \rightarrow FeCl_2 + H_2S$$

The above chemical equation consists of 4 terms: FeS, 2 HCl, $FeCl_2$, and H_2S.

A **coefficient** is a number indicating how much of a substance there is. The coefficient multiplies the number of atoms in a molecule. The coefficient appears to the left of the molecule.

For example, in 7 $Al_2(SO_4)_3$, the number 7 is a coefficient. It indicates that there are 7 aluminum sulfate molecules. The 7 multiplies each atom in the term. (We'll learn more about this concept later in this chapter.)

A chemical reaction is said to be **balanced** when there are the same number of each type of atom on both sides of the chemical equation.

For example, the following chemical equation is balanced:

$$2\ Cu_2S + 3\ O_2 \rightarrow 2\ Cu_2O + 2\ SO_2$$

There are 4 copper (Cu) atoms on each side, 2 sulfur (S) atoms on each side, and 6 oxygen (O) atoms on each side (since $3 \times 2 = 6$ on the left and $2 + 2 \times 2 = 6$ on the right). When every element has the same number of atoms on each side of the reaction, the equation is balanced. (We'll learn how to count atoms later in this chapter.)

According to the **law of conservation of mass**, atoms are neither created nor destroyed during a chemical reaction. This is why chemical reactions need to be balanced: to ensure that the reaction has the same number of each type of atom on both sides of the chemical equation. The law of conservation of mass applies to ordinary chemical reactions.*

* However, the law of conservation of mass doesn't hold true for all types of reactions. For example, in nuclear reactions such as $^{233}U \rightarrow {}^{229}Th + {}^{4}He$, the atoms themselves may change identity.

The Significance of Balancing Reactions

The significance of a balanced chemical equation is that the relative amounts of the reactants and products come in the same proportion as the coefficients in the chemical equation. This way, chemists can prepare the right amount of reactants in the laboratory in order to form the desired amount of products.

For example, consider the following chemical reaction:

$$2\ Na + F_2 \rightarrow 2\ NaF$$

According to the above chemical equation, 2 sodium (Na) atoms react with one diatomic fluorine molecule (F_2) to yield 2 sodium fluoride (NaF) molecules. Because the proportions are the same at any scale, the same chemical equation tells us that 2 moles of sodium react with one mole of diatomic fluorine gas to yield 2 moles of sodium fluoride, for example.

Pre-balancing Skills

Before we learn how to balance a chemical reaction, let us first consider how to count atoms in various terms, using subscripts, parentheses, and coefficients. After all, a chemical equation is basically balanced by counting atoms while inserting coefficients as needed to make the numbers the same on both sides.

A **subscript** denotes the number of atoms present in one molecule. If there is no subscript, there is one atom. When there are two or more atoms, a subscript is used.

Here are a few examples:

- One molecule of CO_2 consists of 1 C atom and 2 O atoms.
- One molecule of Mg_3N_2 consists of 3 Mg atoms and 2 N atoms.
- One molecule of $C_4H_6O_3$ consists of 4 C atoms, 6 H atoms, and 3 O atoms.

When a subscript follows **parentheses**, multiply that subscript by all of the subscripts inside the parentheses.

Following are a few examples:

- One molecule of $Fe_2(SO_4)_3$ consists of 2 Fe atoms, 3 S atoms (since $1 \times 3 = 3$), and 12 O atoms (since $4 \times 3 = 12$).
- One molecule of $Ca(OH)_2$ consists of 1 Ca atom, 2 O atoms (since $1 \times 2 = 2$), and 2 H atoms (since $1 \times 2 = 2$).
- One molecule of $Mg_3(PO_4)_2$ consists of 3 Mg atoms, 2 P atoms (since $1 \times 2 = 2$), and 8 O atoms (since $4 \times 2 = 8$).

A **coefficient** multiplies the numbers of all atoms in the same term. (Recall the words 'coefficient' and 'term' defined earlier in this chapter.) If there is no coefficient, there is one molecule. When there are two or more molecules, a coefficient is used.

Here are a few examples:

- 3 H_2O represents 3 water molecules, consisting of 6 H atoms (since $3 \times 2 = 6$) and 3 O atoms (since $3 \times 1 = 3$).
- 4 N_2O_3 represents 4 dinitrogen trioxide molecules, consisting of 8 N atoms (since $4 \times 2 = 8$) and 12 O atoms (since $4 \times 3 = 12$).
- K_2CO_3 represents 1 potassium carbonate molecule, consisting of 2 K atoms (since $1 \times 2 = 2$), 1 C atom (since $1 \times 1 = 1$), and 3 O atoms (since $1 \times 3 = 3$).
- 2 $C_6H_{12}O_6$ represents 2 glucose molecules, consisting of 12 C atoms (since $2 \times 6 = 12$), 24 H atoms (since $2 \times 12 = 24$), and 12 O atoms (since $2 \times 6 = 12$).

When a coefficient appears in conjunction with a subscript following parentheses, both the coefficient and the subscript multiply all of the subscripts inside the parentheses.

Following are a few examples:

- 4 $Ba(NO_3)_2$ consists of 4 Ba atoms (since $4 \times 1 = 4$), 8 N atoms (since $4 \times 2 \times 1 = 8$), and 24 O atoms (since $4 \times 2 \times 3 = 24$).
- 5 $Fe_2(SO_4)_3$ consists of 10 Fe atoms (since $5 \times 2 = 10$), 15 S atoms (since $5 \times 3 \times 1 = 15$), and 60 O atoms (since $5 \times 3 \times 4 = 60$).
- 2 $(NH_4)_3PO_4$ consists of 6 N atoms (since $2 \times 3 \times 1 = 6$), 24 H atoms (since $2 \times 3 \times 4 = 24$), 2 P atoms (since $2 \times 1 = 2$), and 8 O atoms (since $2 \times 4 = 8$).

When two or more terms are added together (with + signs in between), first treat each term separately and then add all of the atoms together.

Here are a few examples:

- 2 C + O_2 consists of 2 C atoms and 2 O atoms. (The coefficient of C applies only to the first term. Coefficients do not extend past + or → signs.)
- 3 Fe + 4 H_2O consists of 3 Fe atoms, 8 H atoms (since $4 \times 2 = 8$), and 4 O atoms.
- 2 CO_2 + 3 H_2O consists of 2 C atoms, 6 H atoms (since $3 \times 2 = 6$), and 7 O atoms (since $2 \times 2 + 3 \times 1 = 4 + 3 = 7$).

Balancing Strategy (in Words)

Goal: Add coefficients to each term, as needed, such that the total number of each atom on both sides of the chemical equation matches up exactly.

Steps: Follow these guidelines to balance a chemical reaction:

- Begin by counting the number of each type of atom on each side of the chemical equation.
- Insert one coefficient at a time, attempting to balance one element at a time.
- Work with elements that appear in compounds like NaCl or H_2O before working with elements that appear isolated like N_2 or Al.
- Work with elements that appear only once on each side of the equation before dealing with elements that appear two or more times on the same side.
- It's generally better to save H and O for last.
- Realize that you may need to change inserted coefficients over the course of balancing a chemical reaction.
- Use trial and error. It's okay to make a mistake. If something doesn't work out, you can always go back and change it, and try something else instead.
- If you find yourself needing to insert a fractional coefficient at the end, multiply every coefficient by the denominator of that fraction. (See Example 10.)

Check: When you finish balancing a chemical equation, add up the total number of each atom on each side to make sure they are the same on both sides.

Balancing Examples (with Math and Explanations)

The following examples illustrate the strategy for balancing chemical reactions. The examples start out simple and grow progressively more challenging.

Example 1. Balance the following reaction.

$$H_2 + Cl_2 \rightarrow HCl$$

First count atoms on both sides:

- 2 H atoms on the left. 1 H atom on the right.
- 2 Cl atoms on the left. 1 Cl atom on the right.

We need 2 H and 2 Cl atoms on the right, so simply add a coefficient of 2 before the HCl.

The final answer is:

$$H_2 + Cl_2 \rightarrow 2\ HCl$$

(Note that we don't write 1's in front of H_2 or Cl_2. We only write coefficients when they don't equal one.)

Check your answer by counting atoms on both sides:

- 2 H atoms on both sides.
- 2 Cl atoms on both sides.

Example 2. Balance the following reaction.

$$H_2 + O_2 \rightarrow H_2O$$

First count atoms on both sides:

- 2 H atoms on the left. 2 H atoms on the right.
- 2 O atoms on the left. 1 O atom on the right.

Although H is already balanced, O isn't.

We need twice as many O atoms on the right, so insert a coefficient of 2 before H_2O.

$$H_2 + O_2 \rightarrow 2\ H_2O \text{ (unbalanced)}$$

While the coefficient balanced O, it messed up H.

Balance H by adding another coefficient of 2, this time before H_2.

The final answer is:

$$2\ H_2 + O_2 \rightarrow 2\ H_2O$$

Check your answer by counting atoms on both sides:

- 4 H atoms on both sides.
- 2 O atoms on both sides.

Example 3. Balance the following reaction.

$$N_2 + F_2 \rightarrow NF_3$$

First count atoms on both sides:

- 2 N atoms on the left. 1 N atom on the right.
- 2 F atoms on the left. 3 F atoms on the right.

When you see subscripts of 2 and 3 on the same element, it's often helpful to make 6 of each by multiplying 2 by 3 and multiplying 3 by 2 (since 2×3 and 3×2 both equal 6).

Insert a coefficient of 3 before F_2 and a coefficient of 2 before NF_3 to balance F.

In this case, this happens to balance N at the same time. (But things don't always turn out so easy.)

The final answer is:

$$N_2 + 3\ F_2 \rightarrow 2\ NF_3$$

Check your answer by counting atoms on both sides:

- 2 N atoms on both sides.
- 6 F atoms on both sides.

Example 4. Balance the following reaction.

$$Si + O_2 \rightarrow SiO_2$$

First count atoms on both sides:

- 1 Si atom on the left. 1 Si atom on the right.
- 2 O atoms on the left. 2 O atoms on the right.

This is a trick question: It's already balanced.

You don't need to do anything to balance the equation, but you should show your work by counting atoms on each side and stating that it's already balanced (because an instructor might penalize you for leaving an exercise blank).

The final answer is:

$$Si + O_2 \rightarrow SiO_2$$

Example 5. Balance the following reaction.

$$Mg + HCl \rightarrow MgCl_2 + H_2$$

First count atoms on both sides:

- 1 Mg atom on the left. 1 Mg atom on the right.
- 1 H atom on the left. 2 H atoms on the right.
- 1 Cl atom on the left. 2 Cl atoms on the right.

We need 2 H and 2 Cl atoms on the left, so simply add a coefficient of 2 before the HCl.

The final answer is:

$$Mg + 2\ HCl \rightarrow MgCl_2 + H_2$$

Check your answer by counting atoms on both sides:

- 1 Mg atom on both sides.
- 2 H atoms on both sides.
- 2 Cl atoms on both sides.

Example 6. Balance the following reaction.

$$NO_2 + H_2O \rightarrow HNO_3 + NO$$

First count atoms on both sides:

- 1 N atom on the left. 2 N atoms on the right (since N appears in both terms on the right).
- 3 O atoms on the left (since $2 + 1 = 3$). 4 O atoms on the right (since $3 + 1 = 4$).
- 2 H atoms on the left. 1 H atom on the right.

Although the guidelines suggest saving H and O for last, in this case it turns out to be easier to work with O first.

Note that O_2 is on the left and O_3 is on the right. Let's try the trick from Example 3. (That trick doesn't always work, but it's worth testing it out when you see a subscript of 2 on one side and a subscript of 3 on the other side attached to the same element.)

Insert a coefficient of 3 before NO_2 and a coefficient of 2 before HNO_3. The reasoning behind this is that 3×2 and 2×3 both equal 6.

In this case, this happens to also balance N and H at the same time. (But things don't always turn out so easy.)

The final answer is:

$$3\ NO_2 + H_2O \rightarrow 2\ HNO_3 + NO$$

Check your answer by counting atoms on both sides:

- 3 N atoms on both sides. (Note that $2 + 1 = 3$ on the right.)
- 2 H atoms on both sides.
- 7 O atoms on both sides. (Note that $3 \times 2 + 1 = 7$ on the left, while $2 \times 3 + 1 = 7$ on the right.)

Example 7. Balance the following reaction.

$$N_2H_4 + H_2O_2 \rightarrow H_2O + N_2$$

First count atoms on both sides:

- 2 N atoms on the left. 2 N atoms on the right.
- 6 H atoms on the left (since $4 + 2 = 6$). 2 H atoms on the right.
- 2 O atoms on the left. 1 O atom on the right.

Since N is already balanced, we need to deal with H and O first.

According to the guidelines, we should balance O before H since O appears only in one term on each side.

We can balance O by inserting a coefficient of 2 before H_2O.

$$N_2H_4 + H_2O_2 \rightarrow 2\ H_2O + N_2 \text{ (unbalanced)}$$

Although both N and O are now balanced, H isn't.

We still need more H atoms on the right. Changing the coefficient from a 2 to a 3 won't work because we won't be able to make 3 O atoms on the left (since you can't multiply O_2 by an integer to make 3 O atoms).

So let's try changing the coefficient on the right-hand side from a 2 to a 4:

$$N_2H_4 + H_2O_2 \rightarrow 4\ H_2O + N_2 \text{ (unbalanced)}$$

Let's count atoms again to see where we stand:

- 2 N atoms on the left. 2 N atoms on the right.
- 6 H atoms on the left (since $4 + 2 = 6$). 8 H atoms on the right.
- 2 O atoms on the left. 4 O atoms on the right.

We now need 2 more H atoms and 2 more O atoms on the left-hand side. We can achieve this by inserting a coefficient of 2 before H_2O_2.

The final answer is:

$$N_2H_4 + 2\ H_2O_2 \rightarrow 4\ H_2O + N_2$$

Check your answer by counting atoms on both sides:

- 2 N atoms on both sides.
- 8 H atoms on both sides. (Note that $4 + 2 \times 2 = 8$ on the left.)
- 4 O atoms on both sides.

Example 8. Balance the following reaction.

$$Al + Pb(NO_3)_2 \rightarrow Pb + Al(NO_3)_3$$

First count atoms on both sides:

- 1 Al atom on the left. 1 Al atom on the right.
- 1 Pb atom on the left. 1 Pb atom on the right.
- 2 N atoms on the left. 3 N atoms on the right.
- 6 O atoms on the left. 9 O atoms on the right.

According to the guidelines, we should deal with N before O.

With 2 N on the left and 3 N on the right, we can use the trick from Example 3 to balance N: Insert a coefficient of 3 before $Pb(NO_3)_2$ and a coefficient of 2 before $Al(NO_3)_3$. The reasoning behind this is that 2×3 and 3×2 both equal 6.

$$Al + 3\ Pb(NO_3)_2 \rightarrow Pb + 2\ Al(NO_3)_3 \text{ (unbalanced)}$$

Let's count atoms again to see where we stand:

- 1 Al atom on the left. 2 Al atoms on the right.
- 3 Pb atoms on the left. 1 Pb atom on the right.
- 6 N atoms on the left. 6 N atoms on the right. (Note that $3 \times 2 \times 1 = 6$ on the left, while $2 \times 3 \times 1 = 6$ on the right.)
- 18 O atoms on the left. 18 O atoms on the right. (Note that $3 \times 2 \times 3 = 18$ on the left, while $2 \times 3 \times 3 = 18$ on the right.)

Now N and O are both balanced, but Al and Pb are no longer balanced.

We can solve this problem by inserting a coefficient of 2 before Al and a coefficient of 3 before Pb.

The final answer is:

$$2\ Al + 3\ Pb(NO_3)_2 \rightarrow 3\ Pb + 2\ Al(NO_3)_3$$

Check your answer by counting atoms on both sides:

- 2 Al atoms on both sides.
- 3 Pb atoms on both sides.
- 6 N atoms on both sides. (Note that $3 \times 2 \times 1 = 6$ on the left, while $2 \times 3 \times 1 = 6$ on the right.)
- 18 O atoms on both sides. (Note that $3 \times 2 \times 3 = 18$ on the left, while $2 \times 3 \times 3 = 18$ on the right.)

Example 9. Balance the following reaction.

$$Mg_3P_2 + H_2O \rightarrow Mg(OH)_2 + PH_3$$

First count atoms on both sides:

- 3 Mg atoms on the left. 1 Mg atom on the right.
- 2 P atoms on the left. 1 P atom on the right.
- 2 H atoms on the left. 5 H atoms on the right. (Note that $2 + 3 = 5$ on the right.)
- 1 O atom on the left. 2 O atoms on the right.

The guidelines suggest dealing with Mg and P before working with H and O.

Balance Mg by inserting a coefficient of 3 before $Mg(OH)_2$, and balance P by inserting a coefficient of 2 before PH_3.

$$Mg_3P_2 + H_2O \rightarrow 3\ Mg(OH)_2 + 2\ PH_3 \text{ (unbalanced)}$$

Let's count atoms again to see where we stand:

- 3 Mg atoms on the left. 3 Mg atoms on the right.
- 2 P atoms on the left. 2 P atoms on the right.
- 2 H atoms on the left. 12 H atoms on the right. (Note that $3 \times 2 + 2 \times 3 = 6 + 6 = 12$ on the right.)
- 1 O atom on the left. 6 O atoms on the right.

We can balance both H and O by inserting a coefficient of 6 before H_2O.

The final answer is:

$$Mg_3P_2 + 6\ H_2O \rightarrow 3\ Mg(OH)_2 + 2\ PH_3$$

Check your answer by counting atoms on both sides:

- 3 Mg atoms on both sides.
- 2 P atoms on both sides.
- 12 H atoms on both sides.
- 6 O atoms on both sides.

Example 10. Balance the following reaction.

$$C_2H_6 + O_2 \rightarrow CO_2 + H_2O$$

First count atoms on both sides:

- 2 C atoms on the left. 1 C atom on the right.
- 6 H atoms on the left. 2 H atoms on the right.
- 2 O atoms on the left. 3 O atoms on the right.

The guidelines suggest balancing C before dealing with H and O.

Balance C by inserting a 2 before CO_2.

$$C_2H_6 + O_2 \rightarrow 2\ CO_2 + H_2O \text{ (unbalanced)}$$

Work with H before O because O appears in a single-element term, unlike H.

Balance H by inserting a 3 before H_2O.

$$C_2H_6 + O_2 \rightarrow 2\ CO_2 + 3\ H_2O \text{ (unbalanced)}$$

Let's count atoms again to see where we stand:

- 2 C atoms on the left. 2 C atoms on the right.
- 6 H atoms on the left. 6 H atoms on the right.
- 2 O atoms on the left. 7 O atoms on the right. (Note that $2 \times 2 + 3 = 4 + 3 = 7$ on the right.)

Although C and H are balanced, O isn't.

We have a problem: A coefficient of 3.5 before O_2 would balance the reaction (since 3.5 O_2 would make 7 O atoms because $3.5 \times 2 = 7$), but you can't have half a molecule. The following equation would balance the reaction, but that .5 is a problem:

$$C_2H_6 + 3.5\ O_2 \rightarrow 2\ CO_2 + 3\ H_2O \text{ (unfinished)}$$

The solution is to multiply every coefficient by 2 to remove the half:

- C_2H_6 becomes 2 C_2H_6 (since $2 \times 1 = 2$).
- 3.5 O_2 becomes 7 O_2 (since $2 \times 3.5 = 7$).
- 2 CO_2 becomes 4 CO_2 (since $2 \times 2 = 4$).
- 3 becomes 6 H_2O (since $2 \times 3 = 6$).

Now all of the coefficients are integers.

The final answer is:

$$2\ C_2H_6 + 7\ O_2 \rightarrow 4\ CO_2 + 6\ H_2O$$

Check your answer by counting atoms on both sides:

- 4 C atoms on both sides.
- 12 H atoms on both sides.
- 14 O atoms on both sides. (Note that $4 \times 2 + 6 = 8 + 6 = 14$ on the right.)

2 Pre-Balancing Practice

Exercise 1. How many atoms of each kind are there in the following expression?

F_2

Exercise 2. How many atoms of each kind are there in the following expression?

CH_4

Exercise 3. How many atoms of each kind are there in the following expression?

Al_2O_3

Exercise 4. How many atoms of each kind are there in the following expression?

C_2H_5OH

Exercise 5. How many atoms of each kind are there in the following expression?

$Pb(NO_3)_2$

Exercise 6. How many atoms of each kind are there in the following expression?

$Hg_3(PO_4)_2$

Exercise 7. How many atoms of each kind are there in the following expression?

$(NH_4)_2SO_4$

Exercise 8. How many atoms of each kind are there in the following expression?

$3\ N_2$

Exercise 9. How many atoms of each kind are there in the following expression?

$4\ Na_2O$

Exercise 10. How many atoms of each kind are there in the following expression?

$6\ H_2SO_4$

Exercise 11. How many atoms of each kind are there in the following expression?

$$5\ C_{12}H_{22}O_{11}$$

Exercise 12. How many atoms of each kind are there in the following expression?

$$2\ Al_2(CO_3)_3$$

Exercise 13. How many atoms of each kind are there in the following expression?

$$4\ Sn(NO_3)_2$$

Exercise 14. How many atoms of each kind are there in the following expression?

$$5\ (NH_4)_2S$$

Exercise 15. How many atoms of each kind are there in the following expression?

$$2\ Fe + 3\ Cl_2$$

Exercise 16. How many atoms of each kind are there in the following expression?

$$2\ C_6H_{14} + 19\ O_2$$

Exercise 17. How many atoms of each kind are there in the following expression?

$$Al_2(SO_4)_3 + 3\ Ca(OH)_2$$

Exercise 18. How many atoms of each kind are there in the following expression?

$$4Pb(CH_3COO)_2 + 4H_2S$$

3 Basic Structure with 3 Terms or Less

Exercise 1. Balance the following reaction.

$$N_2O_4 \rightarrow NO_2$$

Exercise 2. Balance the following reaction.

$$O_2 \rightarrow O_3$$

Exercise 3. Balance the following reaction.

$$C_6H_6 \rightarrow C_2H_2$$

Exercise 4. Balance the following reaction.

$$C_6H_{12}O_6 \rightarrow CH_2O$$

Exercise 5. Balance the following reaction.

$$C + O_2 \rightarrow CO_2$$

Exercise 6. Balance the following reaction.

$$C + O_2 \rightarrow CO$$

Exercise 7. Balance the following reaction.

$$H_2 + F_2 \rightarrow HF$$

Exercise 8. Balance the following reaction.

$$Xe + F_2 \rightarrow XeF_6$$

Exercise 9. Balance the following reaction.

$$Fe + O_2 \rightarrow FeO$$

Exercise 10. Balance the following reaction.

$$C + F_2 \rightarrow CF_4$$

Exercise 11. Balance the following reaction.

$$U + F_2 \rightarrow UF_6$$

Exercise 12. Balance the following reaction.

$$CH_4 \rightarrow C + H$$

Exercise 13. Balance the following reaction.

$$Ca + O_2 \rightarrow CaO$$

Exercise 14. Balance the following reaction.

$$Mg + N_2 \rightarrow Mg_3N_2$$

Exercise 15. Balance the following reaction.

$$NO + O_2 \rightarrow NO_2$$

Exercise 16. Balance the following reaction.

$$P + Cl_2 \rightarrow PCl_3$$

Exercise 17. Balance the following reaction.

$$H_2 + O_3 \rightarrow H_2O$$

Exercise 18. Balance the following reaction.

$$SO_2 + O_2 \rightarrow SO_3$$

Exercise 19. Balance the following reaction.

$$N_2 + H_2 \rightarrow NH_3$$

Exercise 20. Balance the following reaction.

$$H_2O \rightarrow H_2 + O_2$$

Exercise 21. Balance the following reaction.

$$Co + F_2 \rightarrow CoF_3$$

Exercise 22. Balance the following reaction.

$$Na + Cl_2 \rightarrow NaCl$$

Exercise 23. Balance the following reaction.

$$Fe + C \rightarrow Fe_3C$$

Exercise 24. Balance the following reaction.

$$HgO \rightarrow Hg + O_2$$

Exercise 25. Balance the following reaction.

$$Br_2 + F_2 \rightarrow BrF_3$$

Exercise 26. Balance the following reaction.

$$Mg + O_2 \rightarrow MgO$$

Exercise 27. Balance the following reaction.

$$Al + Cl_2 \rightarrow AlCl_3$$

Exercise 28. Balance the following reaction.

$$SO_3 \rightarrow SO_2 + O_2$$

Exercise 29. Balance the following reaction.

$$Sn + Cl_2 \rightarrow SnCl_4$$

Exercise 30. Balance the following reaction.

$$NaN_3 \rightarrow Na + N_2$$

Exercise 31. Balance the following reaction.

$$P_4 + O_2 \rightarrow P_4O_6$$

Exercise 32. Balance the following reaction.

$$P + F_2 \rightarrow PF_5$$

Exercise 33. Balance the following reaction.

$$N_2O_5 \rightarrow NO_2 + O_2$$

Exercise 34. Balance the following reaction.

$$H_2 + O_2 \rightarrow H_2O$$

Exercise 35. Balance the following reaction.

$$P_4 + O_2 \rightarrow P_4O_{10}$$

Exercise 36. Balance the following reaction.

$$Zn + S_8 \rightarrow ZnS$$

Exercise 37. Balance the following reaction.

$$Br_2 + F_2 \rightarrow BrF_5$$

Exercise 38. Balance the following reaction.

$$FeO + O_2 \rightarrow Fe_2O_3$$

Exercise 39. Balance the following reaction.

$$Fe + Cl_2 \rightarrow FeCl_3$$

Exercise 40. Balance the following reaction.

$$Cr + O_2 \rightarrow Cr_2O_3$$

Exercise 41. Balance the following reaction.

$$Al_2O_3 \rightarrow Al + O_2$$

Exercise 42. Balance the following reaction.

$$PH_3 \rightarrow P_4 + H_2$$

Exercise 43. Balance the following reaction.

$$B + O_2 \rightarrow B_2O_3$$

Exercise 44. Balance the following reaction.

$$Li + N_2 \rightarrow Li_3N$$

Exercise 45. Balance the following reaction.

$$Fe + O_2 \rightarrow Fe_2O_3$$

Exercise 46. Balance the following reaction.

$$P_4 + Br_2 \rightarrow PBr_3$$

Exercise 47. Balance the following reaction.

$$Al + O_2 \rightarrow Al_2O_3$$

Exercise 48. Balance the following reaction.

$$V_2O_5 \rightarrow V + O_2$$

Exercise 49. Balance the following reaction.

$$Rb + S_8 \rightarrow Rb_2S$$

4 Basic Structure with 4 Terms

Exercise 1. Balance the following reaction.

$$Zn + HCl \rightarrow ZnCl_2 + H_2$$

Exercise 2. Balance the following reaction.

$$CH_4 + H_2O \rightarrow CO + H_2$$

Exercise 3. Balance the following reaction.

$$C + H_2O \rightarrow CH_4 + CO_2$$

Exercise 4. Balance the following reaction.

$$KBr + Cl_2 \rightarrow KCl + Br_2$$

Exercise 5. Balance the following reaction.

$$CS_2 + O_2 \rightarrow CO_2 + SO_2$$

Exercise 6. Balance the following reaction.

$$WO_3 + H_2 \rightarrow W + H_2O$$

Exercise 7. Balance the following reaction.

$$CH_4 + O_2 \rightarrow CO_2 + H_2O$$

Exercise 8. Balance the following reaction.

$$CH_4 + Cl_2 \rightarrow CCl_4 + HCl$$

Exercise 9. Balance the following reaction.

$$Al + ZnCl_2 \rightarrow Zn + AlCl_3$$

Exercise 10. Balance the following reaction.

$$CO_2 + H_2 \rightarrow CO + H_2O$$

Exercise 11. Balance the following reaction.

$$CuS + O_2 \rightarrow CuO + SO_2$$

Exercise 12. Balance the following reaction.

$$H_2O + F_2 \rightarrow HF + O_2$$

Exercise 13. Balance the following reaction.

$$TbF_3 + Ca \rightarrow Tb + CaF_2$$

Exercise 14. Balance the following reaction.

$$CoCl_2 + ClF_3 \rightarrow CoF_3 + Cl_2$$

Exercise 15. Balance the following reaction.

$$H_2S + O_2 \rightarrow H_2O + SO_2$$

Exercise 16. Balance the following reaction.

$$HCl + O_2 \rightarrow Cl_2 + H_2O$$

Exercise 17. Balance the following reaction.

$$ZnS + O_2 \rightarrow ZnO + SO_2$$

Exercise 18. Balance the following reaction.

$$Al + HCl \rightarrow AlCl_3 + H_2$$

Exercise 19. Balance the following reaction.

$$F_2 + H_2O \rightarrow HF + O_3$$

Exercise 20. Balance the following reaction.

$$NO_2 + H_2 \rightarrow NH_3 + H_2O$$

Exercise 21. Balance the following reaction.

$$NH_3 + O_2 \rightarrow NO + H_2O$$

Exercise 22. Balance the following reaction.

$$NH_3 + NO \rightarrow N_2 + H_2O$$

Exercise 23. Balance the following reaction.

$$MoS_2 + O_2 \rightarrow MoO_3 + SO_2$$

Exercise 24. Balance the following reaction.

$$PBr_3 + H_2 \rightarrow P_4 + HBr$$

Exercise 25. Balance the following reaction.

$$SO_2 + H_2S \rightarrow S_8 + H_2O$$

Exercise 26. Balance the following reaction.

$$BrF + S_8 \rightarrow SF_4 + Br_2$$

5 Intermediate Structure with 3 Terms

Exercise 1. Balance the following reaction.

$$N_2O + NO_2 \rightarrow NO$$

Exercise 2. Balance the following reaction.

$$C_2H_4 + H_2 \rightarrow C_2H_6$$

Exercise 3. Balance the following reaction.

$$C + H_2 \rightarrow C_5H_{12}$$

Exercise 4. Balance the following reaction.

$$K_2O + H_2O \rightarrow KOH$$

Exercise 5. Balance the following reaction.

$$KClO_3 \rightarrow KCl + O_2$$

Exercise 6. Balance the following reaction.

$$Na_2O + H_2O \rightarrow NaOH$$

Exercise 7. Balance the following reaction.

$$H_2SO_3 + O_2 \rightarrow H_2SO_4$$

Exercise 8. Balance the following reaction.

$$P_4O_{10} + H_2O \rightarrow H_3PO_4$$

Exercise 9. Balance the following reaction.

$$Ni(CO)_4 \rightarrow Ni + CO$$

Exercise 10. Balance the following reaction.

$$CaO + P_2O_5 \rightarrow Ca_3(PO_4)_2$$

Exercise 11. Balance the following reaction.

$$NH_3 + H_2SO_4 \rightarrow (NH_4)_2SO_4$$

Exercise 12. Balance the following reaction.

$$Ca_3(PO_4)_2 + H_3PO_4 \rightarrow Ca(H_2PO_4)_2$$

Exercise 13. Balance the following reaction.

$$CaO + P_4O_{10} \rightarrow Ca_3(PO_4)_2$$

Exercise 14. Balance the following reaction.

$$NH_4NO_3 \rightarrow N_2O + H_2O$$

Exercise 15. Balance the following reaction.

$$Na_2SO_3 + S_8 \rightarrow Na_2S_2O_3$$

Exercise 16. Balance the following reaction.

$$C_6H_{12}O_6 \rightarrow C_2H_5OH + CO_2$$

Exercise 17. Balance the following reaction.

$$C_{12}H_{22}O_{11} \rightarrow C + H_2O$$

6 Intermediate Structure with 4 Terms

Exercise 1. Balance the following reaction.

$$NO_2 + O_3 \rightarrow N_2O_5 + O_2$$

Exercise 2. Balance the following reaction.

$$CS_2 + Cl_2 \rightarrow CCl_4 + S_2Cl_2$$

Exercise 3. Balance the following reaction.

$$Al + Fe_2O_3 \rightarrow Fe + Al_2O_3$$

Exercise 4. Balance the following reaction.

$$UO_2 + HF \rightarrow UF_4 + H_2O$$

Exercise 5. Balance the following reaction.

$$Fe_2O_3 + C \rightarrow Fe + CO$$

Exercise 6. Balance the following reaction.

$$Fe_2O_3 + CO \rightarrow Fe + CO_2$$

Exercise 7. Balance the following reaction.

$$N_2H_4 + N_2O_4 \rightarrow N_2 + H_2O$$

Exercise 8. Balance the following reaction.

$$SiCl_4 + H_2O \rightarrow SiO_2 + HCl$$

Exercise 9. Balance the following reaction.

$$Al_2O_3 + C \rightarrow Al + CO_2$$

Exercise 10. Balance the following reaction.

$$N_2H_4 + NO_2 \rightarrow N_2 + H_2O$$

Exercise 11. Balance the following reaction.

$$C + As_2O_3 \rightarrow CO_2 + As$$

Exercise 12. Balance the following reaction.

$$Fe + H_2O \rightarrow Fe_3O_4 + H_2$$

Exercise 13. Balance the following reaction.

$$I_2O_5 + CO \rightarrow I_2 + CO_2$$

Exercise 14. Balance the following reaction.

$$C_2H_4 + F_2 \rightarrow CF_4 + HF$$

Exercise 15. Balance the following reaction.

$$Fe_2O_3 + S \rightarrow Fe + SO_2$$

Exercise 16. Balance the following reaction.

$$Na + Fe_2O_3 \rightarrow Fe + Na_2O$$

Exercise 17. Balance the following reaction.

$$Cr_2O_3 + Si \rightarrow Cr + SiO_2$$

Exercise 18. Balance the following reaction.

$$C_2F_4 + BrF_3 \rightarrow C_2F_6 + Br_2$$

Exercise 19. Balance the following reaction.

$$Ca + V_2O_5 \rightarrow V + CaO$$

Exercise 20. Balance the following reaction.

$$Fe_2O_3 + Cl_2 \rightarrow FeCl_3 + O_2$$

Exercise 21. Balance the following reaction.

$$Al + Fe_3O_4 \rightarrow Fe + Al_2O_3$$

Exercise 22. Balance the following reaction.

$$Na + H_2O \rightarrow NaOH + H_2$$

Exercise 23. Balance the following reaction.

$$BaCl_2 + H_2SO_4 \rightarrow BaSO_4 + HCl$$

Exercise 24. Balance the following reaction.

$$K + H_2O \rightarrow KOH + H_2$$

Exercise 25. Balance the following reaction.

$$AgNO_3 + KI \rightarrow AgI + KNO_3$$

Exercise 26. Balance the following reaction.

$$KO_2 + H_2O \rightarrow KOH + O_2$$

Exercise 27. Balance the following reaction.

$$Fe + CuSO_4 \rightarrow Cu + Fe_2(SO_4)_3$$

Exercise 28. Balance the following reaction.

$$KI + Pb(NO_3)_2 \rightarrow PbI_2 + KNO_3$$

Exercise 29. Balance the following reaction.

$$Al + H_2SO_4 \rightarrow Al_2(SO_4)_3 + H_2$$

Exercise 30. Balance the following reaction.

$$Fe + H_2O + O_2 \rightarrow Fe(OH)_3$$

Exercise 31. Balance the following reaction.

$$Fe_2O_3 + P \rightarrow Fe + P_4O_{10}$$

Exercise 32. Balance the following reaction.

$$B_2H_6 + O_2 \rightarrow B_2O_3 + H_2O$$

Exercise 33. Balance the following reaction.

$$C_3H_8 + O_2 \rightarrow CO_2 + H_2O$$

Exercise 34. Balance the following reaction.

$$C_2H_6 + O_2 \rightarrow CO_2 + H_2O$$

Exercise 35. Balance the following reaction.

$$C_4H_{10} + O_2 \rightarrow CO_2 + H_2O$$

Exercise 36. Balance the following reaction.

$$Bi_2S_3 + O_2 \rightarrow Bi_2O_3 + SO_2$$

Exercise 37. Balance the following reaction.

$$P_4S_3 + O_2 \rightarrow P_4O_{10} + SO_2$$

Exercise 38. Balance the following reaction.

$$C_2H_2 + O_2 \rightarrow CO_2 + H_2O$$

Exercise 39. Balance the following reaction.

$$C_5H_{12} + O_2 \rightarrow CO_2 + H_2O$$

Exercise 40. Balance the following reaction.

$$C_8H_{18} + O_2 \rightarrow CO_2 + H_2O$$

Exercise 41. Balance the following reaction.

$$B_5H_9 + O_2 \rightarrow B_2O_3 + H_2O$$

Exercise 42. Balance the following reaction.

$$C_6H_{14} + O_2 \rightarrow CO_2 + H_2O$$

Exercise 43. Balance the following reaction.

$$FeS_2 + O_2 \rightarrow Fe_2O_3 + SO_2$$

Exercise 44. Balance the following reaction.

$$C_{12}H_{26} + O_2 \rightarrow CO_2 + H_2O$$

Exercise 45. Balance the following reaction.

$$C_{21}H_{44} + O_2 \rightarrow CO_2 + H_2O$$

Exercise 46. Balance the following reaction.

$$C_{10}H_{22} + O_2 \rightarrow CO_2 + H_2O$$

7 Advanced Structure with 4 Terms

Exercise 1. Balance the following reaction.

$$PBr_3 + H_2O \rightarrow H_3PO_3 + HBr$$

Exercise 2. Balance the following reaction.

$$Al_2O_3 + HCl \rightarrow AlCl_3 + H_2O$$

Exercise 3. Balance the following reaction.

$$PCl_5 + H_2O \rightarrow H_3PO_4 + HCl$$

Exercise 4. Balance the following reaction.

$$SiO_2 + HF \rightarrow H_2SiF_6 + H_2O$$

Exercise 5. Balance the following reaction.

$$PbS + H_2O_2 \rightarrow PbSO_4 + H_2O$$

Exercise 6. Balance the following reaction.

$$Fe_2O_3 + HCl \rightarrow FeCl_3 + H_2O$$

Exercise 7. Balance the following reaction.

$$N_2H_4 + H_2O_2 \rightarrow NO_2 + H_2O$$

Exercise 8. Balance the following reaction.

$$Al_2O_3 + HI \rightarrow AlI_3 + H_2O$$

Exercise 9. Balance the following reaction.

$$Mg(OH)_2 + HNO_3 \rightarrow Mg(NO_3)_2 + H_2O$$

Exercise 10. Balance the following reaction.

$$Al_2(SO_4)_3 + Ca(OH)_2 \rightarrow Al(OH)_3 + CaSO_4$$

Exercise 11. Balance the following reaction.

$$Mg_3N_2 + H_2SO_4 \rightarrow MgSO_4 + (NH_4)_2SO_4$$

Exercise 12. Balance the following reaction.

$$Ca_3(PO_4)_2 + H_2SO_4 \rightarrow CaSO_4 + H_3PO_4$$

Exercise 13. Balance the following reaction.

$$Ca_3P_2 + H_2O \rightarrow Ca(OH)_2 + PH_3$$

Exercise 14. Balance the following reaction.

$$(NH_4)_2Cr_2O_7 \rightarrow Cr_2O_3 + H_2O + N_2$$

Exercise 15. Balance the following reaction.

$$Al_2S_3 + H_2O \rightarrow Al(OH)_3 + H_2S$$

Exercise 16. Balance the following reaction.

$$Al_4C_3 + H_2O \rightarrow Al(OH)_3 + CH_4$$

Exercise 17. Balance the following reaction.

$$Al(OH)_3 + H_2SO_4 \rightarrow Al_2(SO_4)_3 + H_2O$$

Exercise 18. Balance the following reaction.

$$Ba(OH)_2 + CH_3CO_2H \rightarrow Ba\ (CH_3CO_2)_2 + H_2O$$

Exercise 19. Balance the following reaction.

$$Fe(OH)_3 + H_2SO_4 \rightarrow Fe_2(SO_4)_3 + H_2O$$

Exercise 20. Balance the following reaction.

$$Cu + Cu(NH_3)_4Cl_2 + NH_3 \rightarrow Cu(NH_3)_4Cl$$

Exercise 21. Balance the following reaction.

$$Na_3PO_4 + Ba(NO_3)_2 \rightarrow NaNO_3 + Ba_3(PO_4)_2$$

Exercise 22. Balance the following reaction.

$$Pb(CH_3COO)_2 + H_2S \rightarrow PbS + CH_3COOH$$

Exercise 23. Balance the following reaction.

$$CuCl_2 + (NH_4)_3PO_4 \rightarrow Cu_3(PO_4)_2 + NH_4Cl$$

Exercise 24. Balance the following reaction.

$$NH_4NO_3 \rightarrow H_2O + N_2 + O_2$$

Exercise 25. Balance the following reaction.

$$C + H_2 + O_2 \rightarrow C_2H_5OH$$

Exercise 26. Balance the following reaction.

$$C_2H_5OH + O_2 \rightarrow CO_2 + H_2O$$

Exercise 27. Balance the following reaction.

$$C_4H_{10}O + O_2 \rightarrow CO_2 + H_2O$$

Exercise 28. Balance the following reaction.

$$C_6H_{12} + O_2 \rightarrow H_2C_6H_8O_4 + H_2O$$

Exercise 29. Balance the following reaction.

$$C_7H_6O_3 + C_4H_6O_3 \rightarrow C_9H_8O_4 + HC_2H_3O_2$$

Exercise 30. Balance the following reaction.

$$C_7H_6O_3 + C_4H_6O_3 \rightarrow C_9H_8O_4 + H_2O$$

Exercise 31. Balance the following reaction.

$$CH_3OH + O_2 \rightarrow CO_2 + H_2O$$

Exercise 32. Balance the following reaction.

$$C_3H_7COOH + O_2 \rightarrow CO_2 + H_2O$$

Exercise 33. Balance the following reaction.

$$CO_2 + H_2O \rightarrow C_6H_{12}O_6 + O_2$$

Exercise 34. Balance the following reaction.

$$C_5H_{10}O_2 + O_2 \rightarrow CO_2 + H_2O$$

Exercise 35. Balance the following reaction.

$$CO_2 + H_2O \rightarrow C_{12}H_{22}O_{11} + O_2$$

Exercise 36. Balance the following reaction.

$$C_{45}H_{86}O_6 + O_2 \rightarrow CO_2 + H_2O$$

Exercise 37. Balance the following reaction.

$$C_{57}H_{110}O_6 + O_2 \rightarrow CO_2 + H_2O$$

8 Advanced Structure with 5 Terms

Exercise 1. Balance the following reaction.

$$MnO_2 + HCl \rightarrow MnCl_2 + H_2O + Cl_2$$

Exercise 2. Balance the following reaction.

$$Si_3N_4 + CO_2 \rightarrow SiO + N_2O + CO$$

Exercise 3. Balance the following reaction.

$$NF_3 + H_2O \rightarrow HF + NO + NO_2$$

Exercise 4. Balance the following reaction.

$$MnO_2 + HBr \rightarrow MnBr_2 + H_2O + Br_2$$

Exercise 5. Balance the following reaction.

$$FeCl_2 + H_2O \rightarrow Fe_3O_4 + HCl + H_2$$

Exercise 6. Balance the following reaction.

$$Si_3N_4 + CO_2 \rightarrow SiO_2 + N_2O + CO$$

Exercise 7. Balance the following reaction.

$$CH_4 + O_2 + Cl_2 \rightarrow HCl + CO$$

Exercise 8. Balance the following reaction.

$$Cu + H_2SO_4 \rightarrow CuSO_4 + SO_2 + H_2O$$

Exercise 9. Balance the following reaction.

$$C_3H_6 + NH_3 + O_2 \rightarrow C_3H_3N + H_2O$$

Exercise 10. Balance the following reaction.

$$KO_2 + CO_2 + H_2O \rightarrow KHCO_3 + O_2$$

Exercise 11. Balance the following reaction.

$$CuFeS_2 + O_2 \rightarrow Cu + FeO + SO_2$$

Exercise 12. Balance the following reaction.

$$CaSiO_3 + HF \rightarrow CaF_2 + H_2SiF_6 + H_2O$$

Exercise 13. Balance the following reaction.

$$C_3H_6 + NO \rightarrow C_3H_3N + H_2O + N_2$$

Exercise 14. Balance the following reaction.

$$H_2S + HNO_3 \rightarrow NO + H_2O + S$$

Exercise 15. Balance the following reaction.

$$Na_2SiO_3 + HF \rightarrow NaF + H_2SiF_6 + H_2O$$

Exercise 16. Balance the following reaction.

$$NaOH + H_2O_2 + H_2S \rightarrow Na_2SO_4 + H_2O$$

Exercise 17. Balance the following reaction.

$$CH_3NH_2 + H_2O \rightarrow CH_4 + CO_2 + NH_3$$

Exercise 18. Balance the following reaction.

$$HNO_3 + H_2S \rightarrow H_2SO_4 + NO_2 + H_2O$$

Exercise 19. Balance the following reaction.

$$Al + NaOH + H_2O \rightarrow NaAlO_2 + H_2$$

Exercise 20. Balance the following reaction.

$$Cu + HNO_3 \rightarrow Cu(NO_3)_2 + NO_2 + H_2O$$

Exercise 21. Balance the following reaction.

$$Cu + HNO_3 \rightarrow Cu(NO_3)_2 + NO + H_2O$$

Exercise 22. Balance the following reaction.

$$Al + HNO_3 \rightarrow Al(NO_3)_3 + NO + H_2O$$

Exercise 23. Balance the following reaction.

$$NaOH + Al(OH)_3 + HF \rightarrow Na_3AlF_6 + H_2O$$

Exercise 24. Balance the following reaction.

$$Pb + HNO_3 \rightarrow Pb(NO_3)_2 + NO + H_2O$$

Exercise 25. Balance the following reaction.

$$(CH_3)_3N + H_2O \rightarrow CH_4 + CO_2 + NH_3$$

Exercise 26. Balance the following reaction.

$$Al + NaOH + H_2O \rightarrow NaAl(OH)_4 + H_2$$

Exercise 27. Balance the following reaction.

$$FeS_2 + H_2O + O_2 \rightarrow Fe_2(SO_4)_3 + H_2SO_4$$

Exercise 28. Balance the following reaction.

$$NH_4ClO_4 \rightarrow H_2O + N_2 + Cl_2 + O_2$$

Exercise 29. Balance the following reaction.

$$C_6H_{10}O_4 + NH_3 + H_2 \rightarrow C_6H_{16}N_2 + H_2O$$

Exercise 30. Balance the following reaction.

$$C_2H_5NO_2 + O_2 \rightarrow CO_2 + H_2O + N_2$$

Exercise 31. Balance the following reaction.

$$C_7H_5N_3O_6 \rightarrow CO + C + H_2 + N_2$$

Exercise 32. Balance the following reaction.

$$C_3H_5N_3O_9 \rightarrow CO_2 + H_2O + N_2 + O_2$$

Exercise 33. Balance the following reaction.

$$C_8H_{10}N_4O_2 + O_2 \rightarrow CO + H_2O + NO$$

Exercise 34. Balance the following reaction.

$$KClO_3 + C_{12}H_{22}O_{11} \rightarrow KCl + CO_2 + H_2O$$

Exercise 35. Balance the following reaction.

$$C_8H_{10}N_4O_2 + O_2 \rightarrow CO_2 + H_2O + NO_2$$

Exercise 36. Balance the following reaction.

$$NaHCO_3 + H_3C_6H_5O_7 \rightarrow Na_3C_6H_5O_7 + CO_2 + H_2O$$

Exercise 37. Balance the following reaction.

$$ZnCO_3 + C_6H_8O_7 \rightarrow Zn_3(C_6H_5O_7)_2 + CO_2 + H_2O$$

Exercise 38. Balance the following reaction.

$$C_3H_5(NO_3)_3 \rightarrow CO_2 + H_2O + N_2 + O_2$$

Exercise 39. Balance the following reaction.

$$(CH_3)_2NNH_2 + O_2 \rightarrow N_2 + CO_2 + H_2O$$

Exercise 40. Balance the following reaction.

$$Ca_{10}F_2(PO_4)_6 + H_2SO_4 \rightarrow Ca(H_2PO_4)_2 + CaSO_4 + HF$$

9 Advanced Structure with 6 or More Terms

Exercise 1. Balance the following reaction.

$$KNO_3 + C + S_8 \rightarrow K_2S + CO_2 + N_2$$

Exercise 2. Balance the following reaction.

$$Al + NH_4ClO_4 \rightarrow Al_2O_3 + AlCl_3 + NO + H_2O$$

Exercise 3. Balance the following reaction.

$$H_2S + CO_2 + O_2 \rightarrow C_6H_{12}O_6 + S + H_2O$$

Exercise 4. Balance the following reaction.

$$NaCl + NH_3 + CO_2 + H_2O \rightarrow NaHCO_3 + NH_4Cl$$

Exercise 5. Balance the following reaction.

$$KOH + C_6H_5NH_2 + CHCl_3 \rightarrow KCl + C_6H_5NC + H_2O$$

Exercise 6. Balance the following reaction.

$$KMnO_4 + HCl \rightarrow KCl + MnCl_2 + H_2O + Cl_2$$

Exercise 7. Balance the following reaction.

$$NaClO_2 + H_2SO_4 \rightarrow Na_2SO_4 + ClO_2 + HCl + H_2O$$

Exercise 8. Balance the following reaction.

$$NH_4NO_3 + C_{10}H_{22} + O_2 \rightarrow CO_2 + H_2O + N_2$$

Exercise 9. Balance the following reaction.

$$NaN_3 + KNO_3 + SiO_2 \rightarrow Na_4SiO_4 + K_4SiO_4 + N_2$$

Exercise 10. Balance the following reaction.

$$Ca_3(PO_4)_2 + SiO_2 + C \rightarrow CaSiO_3 + CO + P$$

Exercise 11. Balance the following reaction.

$$NaHCO_3 + Ca(H_2PO_4)_2 \rightarrow Na_2HPO_4 + CaHPO_4 + CO_2 + H_2O$$

Exercise 12. Balance the following reaction.

$$(CH_3)_2N_2H_2 + N_2H_4 + N_2O_4 \rightarrow CO_2 + H_2O + N_2$$

Exercise 13. Balance the following reaction.

$$Au + KCN + H_2O + O_2 \rightarrow KAu(CN)_2 + KOH$$

Exercise 14. Balance the following reaction.

$$Ca_3(PO_4)_2 + SiO_2 + C \rightarrow CaSiO_3 + CO + P_4$$

Exercise 15. Balance the following reaction.

$$AgBr + NaOH + C_6H_6O_2 \rightarrow Ag + NaBr + C_6H_4O_2 + H_2O$$

Exercise 16. Balance the following reaction.

$$KMnO_4 + H_2C_2O_4 + H_2SO_4 \rightarrow K_2SO_4 + MnSO_4 + CO_2 + H_2O$$

Exercise 17. Balance the following reaction.

$$K_2CO_3 + CH_4 + H_2O + N_2 + O_2 \rightarrow KHCO_3 + NH_3$$

Exercise 18. Balance the following reaction.

$$FeSO_4 + K_2Cr_2O_7 + H_2SO_4 \rightarrow Fe_2(SO_4)_3 + K_2SO_4 + Cr_2(SO_4)_3 + H_2O$$

Exercise 19. Balance the following reaction.

$$Na_2Cr_2O_7 + C_6H_{10} + H_2SO_4 \rightarrow Na_2SO_4 + Cr_2(SO_4)_3 + C_6H_{10}O_4 + H_2O$$

Answer Key

2 Pre-Balancing Practice

#1) F_2
2 F
#2) CH_4
1 C
4 H
#3) Al_2O_3
2 Al
3 O
#4) C_2H_5OH
2 C
6 H
1 O
#5) $Pb(NO_3)_2$
1 Pb
2 N
6 O
#6) $Hg_3(PO_4)_2$
3 Hg
2 P
8 O
#7) $(NH_4)_2SO_4$
2 N
8 H
1 S
4 O
#8) 3 N_2
6 N
#9) 4 Na_2O
8 Na
4 O
#10) 6 H_2SO_4
12 H
6 S
24 O
#11) 5 $C_{12}H_{22}O_{11}$
60 C
110 H
55 O
#12) 2 $Al_2(CO_3)_3$
4 Al
6 C
18 O
#13) 4 $Sn(NO_3)_2$
4 Sn
8 N
24 O
#14) 5 $(NH_4)_2S$
10 N
40 H
5 S
#15) 2 Fe + 3 Cl_2
2 Fe
6 Cl
#16) 2 C_6H_{14} + 19 O_2
12 C
28 H
38 O
#17) $Al_2(SO_4)_3$ + 3 $Ca(OH)_2$
2 Al
3 S
18 O
3 Ca
6 H
#18) $4Pb(CH_3COO)_2 + 4H_2S$
4 Pb
16 C
32 H
16 O
4 S

3 Basic Structure with 3 Terms or Less

#1) $N_2O_4 \rightarrow 2\ NO_2$

#2) $3\ O_2 \rightarrow 2\ O_3$

#3) $C_6H_6 \rightarrow 3\ C_2H_2$

#4) $C_6H_{12}O_6 \rightarrow 6\ CH_2O$

#5) $C + O_2 \rightarrow CO_2$ (already balanced)

#6) $2\ C + O_2 \rightarrow 2\ CO$

#7) $H_2 + F_2 \rightarrow 2\ HF$

#8) $Xe + 3\ F_2 \rightarrow XeF_6$

#9) $2\ Fe + O_2 \rightarrow 2\ FeO$

#10) $C + 2\ F_2 \rightarrow CF_4$

#11) $U + 3\ F_2 \rightarrow UF_6$

#12) $CH_4 \rightarrow C + 4\ H$

#13) $2\ Ca + O_2 \rightarrow 2\ CaO$

#14) $3\ Mg + N_2 \rightarrow Mg_3N_2$

#15) $2\ NO + O_2 \rightarrow 2\ NO_2$

#16) $2\ P + 3\ Cl_2 \rightarrow 2\ PCl_3$

#17) $3\ H_2 + O_3 \rightarrow 3\ H_2O$

#18) $2\ SO_2 + O_2 \rightarrow 2\ SO_3$

#19) $N_2 + 3\ H_2 \rightarrow 2\ NH_3$

#20) $2\ H_2O \rightarrow 2\ H_2 + O_2$

#21) $2\ Co + 3\ F_2 \rightarrow 2\ CoF_3$

#22) $2\ Na + Cl_2 \rightarrow 2\ NaCl$

#23) $3\ Fe + C \rightarrow Fe_3C$

#24) $2\ HgO \rightarrow 2\ Hg + O_2$

#25) $Br_2 + 3\ F_2 \rightarrow 2\ BrF_3$

#26) $2\ Mg + O_2 \rightarrow 2\ MgO$

#27) $2\ Al + 3\ Cl_2 \rightarrow 2\ AlCl_3$

#28) $2\ SO_3 \rightarrow 2\ SO_2 + O_2$

#29) $Sn + 2\ Cl_2 \rightarrow SnCl_4$

#30) $2\ NaN_3 \rightarrow 2\ Na + 3\ N_2$

#31) $P_4 + 3\ O_2 \rightarrow P_4O_6$

#32) $2\ P + 5\ F_2 \rightarrow 2\ PF_5$

#33) $2\ N_2O_5 \rightarrow 4\ NO_2 + O_2$

#34) $2\ H_2 + O_2 \rightarrow 2\ H_2O$

#35) $P_4 + 5\ O_2 \rightarrow P_4O_{10}$

#36) $8\ Zn + S_8 \rightarrow 8\ ZnS$

#37) $Br_2 + 5\ F_2 \rightarrow 2\ BrF_5$

#38) $4\ FeO + O_2 \rightarrow 2\ Fe_2O_3$

#39) $2\ Fe + 3\ Cl_2 \rightarrow 2\ FeCl_3$

#40) $4\ Cr + 3\ O_2 \rightarrow 2\ Cr_2O_3$

#41) $2\ Al_2O_3 \rightarrow 4\ Al + 3\ O_2$

#42) $4\ PH_3 \rightarrow P_4 + 6\ H_2$

#43) $4\ B + 3\ O_2 \rightarrow 2\ B_2O_3$

#44) $6\ Li + N_2 \rightarrow 2\ Li_3N$

#45) $4\ Fe + 3\ O_2 \rightarrow 2\ Fe_2O_3$

#46) $P_4 + 6\ Br_2 \rightarrow 4\ PBr_3$

#47) $4\ Al + 3\ O_2 \rightarrow 2\ Al_2O_3$

#48) $2\ V_2O_5 \rightarrow 4\ V + 5\ O_2$

#49) $16\ Rb + S_8 \rightarrow 8\ Rb_2S$

4 Basic Structure with 4 Terms

#1) $Zn + 2\ HCl \rightarrow ZnCl_2 + H_2$

#2) $CH_4 + H_2O \rightarrow CO + 3\ H_2$

#3) $2\ C + 2\ H_2O \rightarrow CH_4 + CO_2$

#4) $2\ KBr + Cl_2 \rightarrow 2\ KCl + Br_2$

#5) $CS_2 + 3\ O_2 \rightarrow CO_2 + 2\ SO_2$

#6) $WO_3 + 3\ H_2 \rightarrow W + 3\ H_2O$

#7) $CH_4 + 2\ O_2 \rightarrow CO_2 + 2\ H_2O$

#8) $CH_4 + 4\ Cl_2 \rightarrow CCl_4 + 4\ HCl$

#9) $2\ Al + 3\ ZnCl_2 \rightarrow 3\ Zn + 2\ AlCl_3$

#10) $CO_2 + H_2 \rightarrow CO + H_2O$ (already balanced)

#11) $2\ CuS + 3\ O_2 \rightarrow 2\ CuO + 2\ SO_2$

#12) $2\ H_2O + 2\ F_2 \rightarrow 4\ HF + O_2$

#13) $2\ TbF_3 + 3\ Ca \rightarrow 2\ Tb + 3\ CaF_2$

#14) $2\ CoCl_2 + 2\ ClF_3 \rightarrow 2\ CoF_3 + 3\ Cl_2$

#15) $2\ H_2S + 3\ O_2 \rightarrow 2\ H_2O + 2\ SO_2$

#16) $4\ HCl + O_2 \rightarrow 2\ Cl_2 + 2\ H_2O$

#17) $2\ ZnS + 3\ O_2 \rightarrow 2\ ZnO + 2\ SO_2$

#18) $2\ Al + 6\ HCl \rightarrow 2\ AlCl_3 + 3\ H_2$

#19) $3\ F_2 + 3\ H_2O \rightarrow 6\ HF + O_3$

#20) $2\ NO_2 + 7\ H_2 \rightarrow 2\ NH_3 + 4\ H_2O$

#21) $4\ NH_3 + 5\ O_2 \rightarrow 4\ NO + 6\ H_2O$

#22) $4\ NH_3 + 6\ NO \rightarrow 5\ N_2 + 6\ H_2O$

#23) $2\ MoS_2 + 7\ O_2 \rightarrow 2\ MoO_3 + 4\ SO_2$

#24) $4\ PBr_3 + 6\ H_2 \rightarrow P_4 + 12\ HBr$

#25) $8\ SO_2 + 16\ H_2S \rightarrow 3\ S_8 + 16\ H_2O$

#26) $32\ BrF + S_8 \rightarrow 8\ SF_4 + 16\ Br_2$

5 Intermediate Structure with 3 Terms

#1) $N_2O + NO_2 \rightarrow 3\ NO$

#2) $C_2H_4 + H_2 \rightarrow C_2H_6$ (already balanced)

#3) $5\ C + 6\ H_2 \rightarrow C_5H_{12}$

#4) $K_2O + H_2O \rightarrow 2\ KOH$

#5) $2\ KClO_3 \rightarrow 2\ KCl + 3\ O_2$

#6) $Na_2O + H_2O \rightarrow 2\ NaOH$

#7) $2\ H_2SO_3 + O_2 \rightarrow 2\ H_2SO_4$

#8) $P_4O_{10} + 6\ H_2O \rightarrow 4\ H_3PO_4$

#9) $Ni(CO)_4 \rightarrow Ni + 4\ CO$

#10) $3\ CaO + P_2O_5 \rightarrow Ca_3(PO_4)_2$

#11) $2\ NH_3 + H_2SO_4 \rightarrow (NH_4)_2SO_4$

#12) $Ca_3(PO_4)_2 + 4\ H_3PO_4 \rightarrow 3\ Ca(H_2PO_4)_2$

#13) $6\ CaO + P_4O_{10} \rightarrow 2\ Ca_3(PO_4)_2$

#14) $NH_4NO_3 \rightarrow N_2O + 2\ H_2O$

#15) $8\ Na_2SO_3 + S_8 \rightarrow 8\ Na_2S_2O_3$

#16) $C_6H_{12}O_6 \rightarrow 2\ C_2H_5OH + 2\ CO_2$

#17) $C_{12}H_{22}O_{11} \rightarrow 12\ C + 11\ H_2O$

6 Intermediate Structure with 4 Terms

#1) $2\ NO_2 + O_3 \rightarrow N_2O_5 + O_2$

#2) $CS_2 + 3\ Cl_2 \rightarrow CCl_4 + S_2Cl_2$

#3) $2\ Al + Fe_2O_3 \rightarrow 2\ Fe + Al_2O_3$

#4) $UO_2 + 4\ HF \rightarrow UF_4 + 2\ H_2O$

#5) $Fe_2O_3 + 3\ C \rightarrow 2\ Fe + 3\ CO$

#6) $Fe_2O_3 + 3\ CO \rightarrow 2\ Fe + 3\ CO_2$

#7) $2\ N_2H_4 + N_2O_4 \rightarrow 3\ N_2 + 4\ H_2O$

#8) $SiCl_4 + 2\ H_2O \rightarrow SiO_2 + 4\ HCl$

#9) $2\ Al_2O_3 + 3\ C \rightarrow 4\ Al + 3\ CO_2$

#10) $2\ N_2H_4 + 2NO_2 \rightarrow 3\ N_2 + 4\ H_2O$

#11) $3\ C + 2\ As_2O_3 \rightarrow 3\ CO_2 + 4\ As$

#12) $3\ Fe + 4\ H_2O \rightarrow Fe_3\ O_4 + 4\ H_2$

#13) $I_2O_5 + 5\ CO \rightarrow I_2 + 5\ CO_2$

#14) $C_2H_4 + 6\ F_2 \rightarrow 2\ CF_4 + 4\ HF$

#15) $2\ Fe_2O_3 + 3\ S \rightarrow 4\ Fe + 3\ SO_2$

#16) $6\ Na + Fe_2O_3 \rightarrow 2\ Fe + 3\ Na_2O$

#17) $2\ Cr_2O_3 + 3\ Si \rightarrow 4\ Cr + 3\ SiO_2$

#18) $3\ C_2F_4 + 2\ BrF_3 \rightarrow 3\ C_2F_6 + Br_2$

#19) $5\ Ca + V_2O_5 \rightarrow 2\ V + 5\ CaO$

#20) $2\ Fe_2O_3 + 6\ Cl_2 \rightarrow 4\ FeCl_3 + 3\ O_2$

#21) $8\ Al + 3\ Fe_3O_4 \rightarrow 9\ Fe + 4\ Al_2O_3$

#22) $2\ Na + 2\ H_2O \rightarrow 2\ NaOH + H_2$

#23) $BaCl_2 + H_2SO_4 \rightarrow BaSO_4 + 2\ HCl$

#24) $2\ K + 2\ H_2O \rightarrow 2\ KOH + H_2$

#25) $AgNO_3 + KI \rightarrow AgI + KNO_3$ (already balanced)

#26) $4\ KO_2 + 2\ H_2O \rightarrow 4\ KOH + 3\ O_2$

#27) $2\ Fe + 3\ CuSO_4 \rightarrow 3\ Cu + Fe_2(SO_4)_3$

#28) $2\ KI + Pb(NO_3)_2 \rightarrow PbI_2 + 2\ KNO_3$

#29) $2\ Al + 3\ H_2SO_4 \rightarrow Al_2(SO_4)_3 + 3\ H_2$

#30) $4\ Fe + 6\ H_2O + 3\ O_2 \rightarrow 4\ Fe(OH)_3$

#31) $10\ Fe_2O_3 + 12\ P \rightarrow 20\ Fe + 3\ P_4O_{10}$

#32) $B_2H_6 + 3\ O_2 \rightarrow B_2O_3 + 3\ H_2O$

#33) $C_3H_8 + 5\ O_2 \rightarrow 3\ CO_2 + 4\ H_2O$

#34) $2\ C_2H_6 + 7\ O_2 \rightarrow 4\ CO_2 + 6\ H_2O$

#35) $2\ C_4H_{10} + 13\ O_2 \rightarrow 8\ CO_2 + 10\ H_2O$

#36) $2\ Bi_2S_3 + 9\ O_2 \rightarrow 2\ Bi_2O_3 + 6\ SO_2$

#37) $P_4S_3 + 8\ O_2 \rightarrow P_4O_{10} + 3\ SO_2$

#38) $2\ C_2H_2 + 5\ O_2 \rightarrow 4\ CO_2 + 2\ H_2O$

#39) $C_5H_{12} + 8\ O_2 \rightarrow 5\ CO_2 + 6\ H_2O$

#40) $2\ C_8H_{18} + 25\ O_2 \rightarrow 16\ CO_2 + 18\ H_2O$

#41) $2\ B_5H_9 + 12\ O_2 \rightarrow 5\ B_2O_3 + 9\ H_2O$

#42) $2\ C_6H_{14} + 19\ O_2 \rightarrow 12\ CO_2 + 14\ H_2O$

#43) $4\ FeS_2 + 11\ O_2 \rightarrow 2\ Fe_2O_3 + 8\ SO_2$

#44) $2\ C_{12}H_{26} + 37\ O_2 \rightarrow 24\ CO_2 + 26\ H_2O$

#45) $C_{21}H_{44} + 32\ O_2 \rightarrow 21\ CO_2 + 22\ H_2O$

#46) $2\ C_{10}H_{22} + 31\ O_2 \rightarrow 20\ CO_2 + 22\ H_2O$

7 Advanced Structure with 4 Terms

#1) $PBr_3 + 3\ H_2O \rightarrow H_3PO_3 + 3\ HBr$

#2) $Al_2O_3 + 6\ HCl \rightarrow 2\ AlCl_3 + 3\ H_2O$

#3) $PCl_5 + 4\ H_2O \rightarrow H_3PO_4 + 5\ HCl$

#4) $SiO_2 + 6\ HF \rightarrow H_2SiF_6 + 2\ H_2O$

#5) $PbS + 4\ H_2O_2 \rightarrow PbSO_4 + 4\ H_2O$

#6) $Fe_2O_3 + 6\ HCl \rightarrow 2\ FeCl_3 + 3\ H_2O$

#7) $N_2H_4 + 6\ H_2O_2 \rightarrow 2\ NO_2 + 8\ H_2O$

#8) $Al_2O_3 + 6\ HI \rightarrow 2\ AlI_3 + 3\ H_2O$

#9) $Mg(OH)_2 + 2\ HNO_3 \rightarrow Mg(NO_3)_2 + 2\ H_2O$

#10) $Al_2(SO_4)_3 + 3\ Ca(OH)_2 \rightarrow 2\ Al(OH)_3 + 3\ CaSO_4$

#11) $Mg_3N_2 + 4\ H_2SO_4 \rightarrow 3\ MgSO_4 + (NH_4)_2SO_4$

#12) $Ca_3(PO_4)_2 + 3\ H_2SO_4 \rightarrow 3\ CaSO_4 + 2\ H_3PO_4$

#13) $Ca_3P_2 + 6\ H_2O \rightarrow 3\ Ca(OH)_2 + 2\ PH_3$

#14) $(NH_4)_2Cr_2O_7 \rightarrow Cr_2O_3 + 4\ H_2O + N_2$

#15) $Al_2S_3 + 6\ H_2O \rightarrow 2\ Al(OH)_3 + 3\ H_2S$

#16) $Al_4C_3 + 12\ H_2O \rightarrow 4\ Al(OH)_3 + 3\ CH_4$

#17) $2\ Al(OH)_3 + 3\ H_2SO_4 \rightarrow Al_2(SO_4)_3 + 6\ H_2O$

#18) $Ba(OH)_2 + 2\ CH_3CO_2H \rightarrow Ba\ (CH_3CO_2)_2 + 2\ H_2O$

#19) $2\ Fe(OH)_3 + 3\ H_2SO_4 \rightarrow Fe_2(SO_4)_3 + 6\ H_2O$

#20) $Cu + Cu(NH_3)_4Cl_2 + 4\ NH_3 \rightarrow 2\ Cu(NH_3)_4Cl$

#21) $2\ Na_3PO_4 + 3\ Ba(NO_3)_2 \rightarrow 6\ NaNO_3 + Ba_3(PO_4)_2$

#22) $Pb(CH_3COO)_2 + H_2S \rightarrow PbS + 2\ CH_3COOH$

#23) $3\ CuCl_2 + 2\ (NH_4)_3PO_4 \rightarrow Cu_3(PO_4)_2 + 6\ NH_4Cl$

#24) $2\ NH_4NO_3 \rightarrow 4\ H_2O + 2\ N_2 + O_2$

#25) $4\ C + 6\ H_2 + O_2 \rightarrow 2\ C_2H_5OH$

#26) $C_2H_5OH + 3\ O_2 \rightarrow 2\ CO_2 + 3\ H_2O$

#27) $C_4H_{10}O + 6\ O_2 \rightarrow 4\ CO_2 + 5\ H_2O$

#28) $2\ C_6H_{12} + 5\ O_2 \rightarrow 2\ H_2C_6H_8O_4 + 2\ H_2O$

#29) $C_7H_6O_3 + C_4H_6O_3 \rightarrow C_9H_8O_4 + HC_2H_3O_2$ (already balanced)

#30) $2\ C_7H_6O_3 + C_4H_6O_3 \rightarrow 2\ C_9H_8O_4 + H_2O$

#31) $2\ CH_3OH + 3\ O_2 \rightarrow 2\ CO_2 + 4\ H_2O$

#32) $C_3H_7COOH + 5\ O_2 \rightarrow 4\ CO_2 + 4\ H_2O$

#33) $6\ CO_2 + 6\ H_2O \rightarrow C_6H_{12}O_6 + 6\ O_2$

#34) $2\ C_5H_{10}O_2 + 13\ O_2 \rightarrow 10\ CO_2 + 10\ H_2O$

#35) $12\ CO_2 + 11\ H_2O \rightarrow C_{12}H_{22}O_{11} + 12\ O_2$

#36) $2\ C_{45}H_{86}O_6 + 127\ O_2 \rightarrow 90\ CO_2 + 86\ H_2O$

#37) $2\ C_{57}H_{110}O_6 + 163\ O_2 \rightarrow 114\ CO_2 + 110\ H_2O$

8 Advanced Structure with 5 Terms

#1) $MnO_2 + 4\ HCl \rightarrow MnCl_2 + 2\ H_2O + Cl_2$

#2) $Si_3N_4 + 5\ CO_2 \rightarrow 3\ SiO + 2\ N_2O + 5\ CO$

#3) $2\ NF_3 + 3\ H_2O \rightarrow 6\ HF + NO + NO_2$

#4) $MnO_2 + 4\ HBr \rightarrow MnBr_2 + 2\ H_2O + Br_2$

#5) $3\ FeCl_2 + 4\ H_2O \rightarrow Fe_3O_4 + 6\ HCl + H_2$

#6) $Si_3N_4 + 8\ CO_2 \rightarrow 3\ SiO_2 + 2\ N_2O + 8\ CO$

#7) $2\ CH_4 + O_2 + 4\ Cl_2 \rightarrow 8\ HCl + 2\ CO$

#8) $Cu + 2\ H_2SO_4 \rightarrow CuSO_4 + SO_2 + 2\ H_2O$

#9) $2\ C_3H_6 + 2\ NH_3 + 3\ O_2 \rightarrow 2\ C_3H_3N + 6\ H_2O$

#10) $4\ KO_2 + 4\ CO_2 + 2\ H_2O \rightarrow 4\ KHCO_3 + 3\ O_2$

#11) $2\ CuFeS_2 + 5\ O_2 \rightarrow 2\ Cu + 2\ FeO + 4\ SO_2$

#12) $CaSiO_3 + 8\ HF \rightarrow CaF_2 + H_2SiF_6 + 3\ H_2O$

#13) $4\ C_3H_6 + 6\ NO \rightarrow 4\ C_3H_3N + 6\ H_2O + N_2$

#14) $3\ H_2S + 2\ HNO_3 \rightarrow 2\ NO + 4\ H_2O + 3\ S$

#15) $Na_2SiO_3 + 8\ HF \rightarrow 2\ NaF + H_2SiF_6 + 3\ H_2O$

#16) $2\ NaOH + 4\ H_2O_2 + H_2S \rightarrow Na_2SO_4 + 6\ H_2O$

#17) $4\ CH_3NH_2 + 2\ H_2O \rightarrow 3\ CH_4 + CO_2 + 4\ NH_3$

#18) $8\ HNO_3 + H_2S \rightarrow H_2SO_4 + 8\ NO_2 + 4\ H_2O$

#19) $2\ Al + 2\ NaOH + 2\ H_2O \rightarrow 2\ NaAlO_2 + 3\ H_2$

#20) $Cu + 4\ HNO_3 \rightarrow Cu(NO_3)_2 + 2\ NO_2 + 2\ H_2O$

#21) $3\ Cu + 8\ HNO_3 \rightarrow 3\ Cu(NO_3)_2 + 2\ NO + 4\ H_2O$

#22) $Al + 4\ HNO_3 \rightarrow Al(NO_3)_3 + NO + 2\ H_2O$

#23) $3\ NaOH + Al(OH)_3 + 6\ HF \rightarrow Na_3AlF_6 + 6\ H_2O$

#24) $3\ Pb + 8\ HNO_3 \rightarrow 3\ Pb(NO_3)_2 + 2\ NO + 4\ H_2O$

#25) $4\ (CH_3)_3N + 6\ H_2O \rightarrow 9\ CH_4 + 3\ CO_2 + 4\ NH_3$

#26) $2\ Al + 2\ NaOH + 6\ H_2O \rightarrow 2\ NaAl(OH)_4 + 3\ H_2$

#27) $4\ FeS_2 + 2\ H_2O + 15\ O_2 \rightarrow 2\ Fe_2(SO_4)_3 + 2\ H_2SO_4$

#28) $2\ NH_4ClO_4 \rightarrow 4\ H_2O + N_2 + Cl_2 + 2\ O_2$

#29) $C_6H_{10}O_4 + 2\ NH_3 + 4\ H_2 \rightarrow C_6H_{16}N_2 + 4\ H_2O$

#30) $4\ C_2H_5NO_2 + 9\ O_2 \rightarrow 8\ CO_2 + 10\ H_2O + 2\ N_2$

#31) $2\ C_7H_5N_3O_6 \rightarrow 12\ CO + 2\ C + 5\ H_2 + 3\ N_2$

#32) $4\ C_3H_5N_3O_9 \rightarrow 12\ CO_2 + 10\ H_2O + 6\ N_2 + O_2$

#33) $2\ C_8H_{10}N_4O_2 + 15\ O_2 \rightarrow 16\ CO + 10\ H_2O + 8\ NO$

#34) $8\ KClO_3 + C_{12}H_{22}O_{11} \rightarrow 8\ KCl + 12\ CO_2 + 11\ H_2O$

#35) $2\ C_8H_{10}N_4O_2 + 27\ O_2 \rightarrow 16\ CO_2 + 10\ H_2O + 8\ NO_2$

#36) $3\ NaHCO_3 + H_3C_6H_5O_7 \rightarrow Na_3C_6H_5O_7 + 3\ CO_2 + 3\ H_2O$

#37) $3\ ZnCO_3 + 2\ C_6H_8O_7 \rightarrow Zn_3(C_6H_5O_7)_2 + 3\ CO_2 + 3\ H_2O$

#38) $4\ C_3H_5(NO_3)_3 \rightarrow 12\ CO_2 + 10\ H_2O + 6\ N_2 + O_2$

#39) $(CH_3)_2NNH_2 + 4\ O_2 \rightarrow N_2 + 2\ CO_2 + 4\ H_2O$

#40) $Ca_{10}F_2(PO_4)_6 + 7\ H_2SO_4 \rightarrow 3\ Ca(H_2PO_4)_2 + 7\ CaSO_4 + 2\ HF$

9 Advanced Structure with 6 or More Terms

#1) $16\ KNO_3 + 24\ C + S_8 \rightarrow 8\ K_2S + 24\ CO_2 + 8\ N_2$

#2) $3\ Al + 3\ NH_4ClO_4 \rightarrow Al_2O_3 + AlCl_3 + 3\ NO + 6\ H_2O$

#3) $24\ H_2S + 6\ CO_2 + 6\ O_2 \rightarrow C_6H_{12}O_6 + 24\ S + 18\ H_2O$

#4) $NaCl + NH_3 + CO_2 + H_2O \rightarrow NaHCO_3 + NH_4Cl$ (already balanced)

#5) $3\ KOH + C_6H_5NH_2 + CHCl_3 \rightarrow 3\ KCl + C_6H_5NC + 3\ H_2O$

#6) $2\ KMnO_4 + 16\ HCl \rightarrow 2\ KCl + 2\ MnCl_2 + 8\ H_2O + 5\ Cl_2$

#7) $10\ NaClO_2 + 5\ H_2SO_4 \rightarrow 5\ Na_2SO_4 + 8\ ClO_2 + 2\ HCl + 4\ H_2O$

#8) $3\ NH_4NO_3 + C_{10}H_{22} + 14\ O_2 \rightarrow 10\ CO_2 + 17\ H_2O + 3\ N_2$

#9) $20\ NaN_3 + 4\ KNO_3 + 6\ SiO_2 \rightarrow 5\ Na_4SiO_4 + K_4SiO_4 + 32\ N_2$

#10) $Ca_3(PO_4)_2 + 3\ SiO_2 + 5\ C \rightarrow 3\ CaSiO_3 + 5\ CO + 2\ P$

#11) $2\ NaHCO_3 + Ca(H_2PO_4)_2 \rightarrow Na_2HPO_4 + CaHPO_4 + 2\ CO_2 + 2\ H_2O$

#12) $(CH_3)_2N_2H_2 + 2\ N_2H_4 + 3\ N_2O_4 \rightarrow 2\ CO_2 + 8\ H_2O + 6\ N_2$

#13) $4\ Au + 8\ KCN + 2\ H_2O + O_2 \rightarrow 4\ KAu(CN)_2 + 4\ KOH$

#14) $2\ Ca_3(PO_4)_2 + 6\ SiO_2 + 10\ C \rightarrow 6\ CaSiO_3 + 10\ CO + P_4$

#15) $2\ AgBr + 2\ NaOH + C_6H_6O_2 \rightarrow 2\ Ag + 2\ NaBr + C_6H_4O_2 + 2\ H_2O$

#16) $2\ KMnO_4 + 5\ H_2C_2O_4 + 3\ H_2SO_4 \rightarrow K_2SO_4 + 2\ MnSO_4 + 10\ CO_2 + 8\ H_2O$

#17) $7\ K_2CO_3 + 7\ CH_4 + 17\ H_2O + 8\ N_2 + 2\ O_2 \rightarrow 14\ KHCO_3 + 16\ NH_3$

#18) $6\ FeSO_4 + K_2Cr_2O_7 + 7\ H_2SO_4 \rightarrow 3\ Fe_2(SO_4)_3 + K_2SO_4 + Cr_2(SO_4)_3 + 7\ H_2O$

#19) $4\ Na_2Cr_2O_7 + 3\ C_6H_{10} + 16\ H_2SO_4 \rightarrow 4\ Na_2SO_4 + 4\ Cr_2(SO_4)_3 + 3\ C_6H_{10}O_4 + 16\ H_2O$

About the Author

Chris McMullen is a physics instructor at Northwestern State University of Louisiana and also an author of academic books. Whether in the classroom or as a writer, Dr. McMullen loves sharing knowledge and the art of motivating and engaging students.

He earned his Ph.D. in phenomenological high-energy physics (particle physics) from Oklahoma State University in 2002. Originally from California, Dr. McMullen earned his Master's degree from California State University, Northridge, where his thesis was in the field of electron spin resonance.

As a physics teacher, Dr. McMullen observed that many students lack fluency in fundamental math skills. In an effort to help students of all ages and levels master basic math skills, he published a series of math workbooks on arithmetic, fractions, and algebra called the Improve Your Math Fluency Series. Dr. McMullen has also published a variety of science books, including introductions to basic astronomy and chemistry concepts in addition to physics textbooks.

Dr. McMullen is very passionate about teaching. Many students and observers have been impressed with the transformation that occurs when he walks into the classroom, and the interactive engaged discussions that he leads during class time. Dr. McMullen is well-known for drawing monkeys and using them in his physics examples and problems, applying his creativity to inspire students. A stressed-out student is likely to be told to throw some bananas at monkeys, smile, and think happy physics thoughts.

Author, Chris McMullen, Ph.D.

Improve Your Math Fluency

This series of math workbooks is geared toward practicing essential math skills:

- Algebra and trigonometry
- Fractions, decimals, and percents
- Long division
- Multiplication and division
- Addition and subtraction

www.improveyourmathfluency.com

www.chrismcmullen.com

Dr. McMullen has published a variety of **science** books, including:

- Basic astronomy concepts
- Basic chemistry concepts
- Creative physics problems
- Calculus-based physics

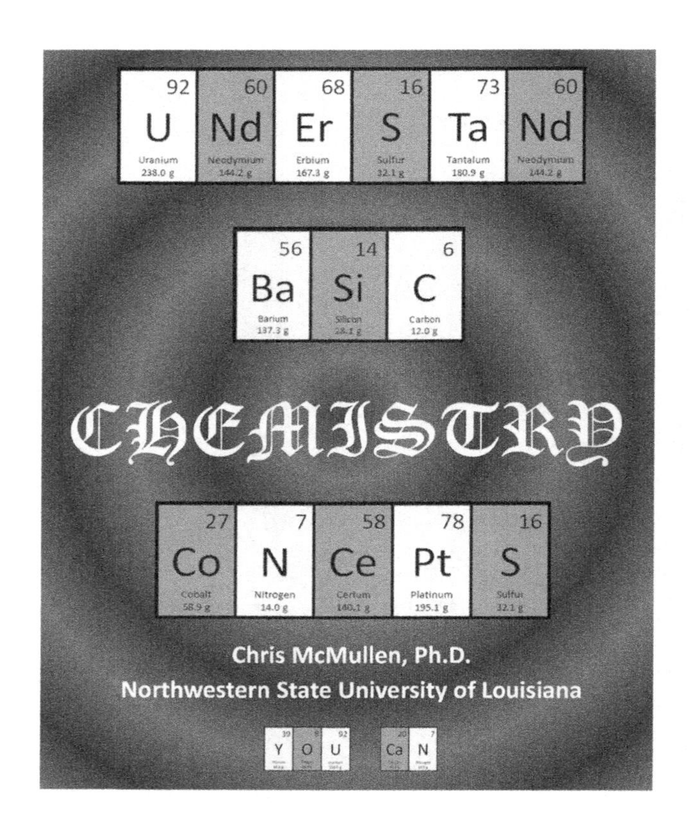

Chris McMullen enjoys solving puzzles. His favorite puzzle is Kakuro (kind of like a cross between crossword puzzles and Sudoku). He once taught a three-week summer course on puzzles.

If you enjoy mathematical pattern puzzles, you might appreciate:

300+ Mathematical Pattern Puzzles

Number Pattern Recognition & Reasoning

- pattern recognition
- visual discrimination
- analytical skills
- logic and reasoning
- analogies
- mathematics

Chris McMullen has coauthored several word scramble books. This includes a cool idea called **VErBAl ReAcTiONS**. A VErBAl ReAcTiON expresses word scrambles so that they look like chemical reactions. Here is an example:

$$2\,C + U + 2\,S + Es \rightarrow S\,U\,C\,C\,Es\,S$$

The left side of the reaction indicates that the answer has 2 C's, 1 U, 2 S's, and 1 Es. Rearrange CCUSSEs to form SUCCEsS.

Each answer to a **VErBAl ReAcTiON** is not merely a word, it's a chemical word. A chemical word is made up not of letters, but of elements of the periodic table. In this case, SUCCEsS is made up of sulfur (S), uranium (U), carbon (C), and Einsteinium (Es).

Another example of a chemical word is GeNiUS. It's made up of germanium (Ge), nickel (Ni), uranium (U), and sulfur (S).

If you enjoy anagrams and like science or math, these puzzles are tailor-made for you.

PERIODIC

1 H hydrogen 1.0 g								
3 Li lithium 6.9 g	4 Be beryllium 9.0 g							
11 Na sodium 23.0 g	12 Mg magnesium 24.3 g							
19 K potassium 39.1 g	20 Ca calcium 40.1 g	21 Sc scandium 45.0 g	22 Ti titanium 47.9 g	23 V vanadium 50.9 g	24 Cr chromium 52.0 g	25 Mn manganese 54.9 g	26 Fe iron 55.8 g	27 Co cobalt 58.9 g
37 Rb rubidium 85.5 g	38 Sr strontium 87.6 g	39 Y yttrium 88.9 g	40 Zr zirconium 91.2 g	41 Nb niobium 92.9 g	42 Mo molybdenum 95.9 g	43 Tc technetium 97.9 g	44 Ru ruthenium 101.1 g	45 Rh rhodium 102.9 g
55 Cs cesium 132.9 g	56 Ba barium 137.3 g	57 La lanthanum 138.9 g	72 Hf hafnium 178.5 g	73 Ta tantalum 180.9 g	74 W tungsten 183.8 g	75 Re rhenium 186.2 g	76 Os osmium 190.2 g	77 Ir iridium 192.2 g
87 Fr francium 223.0 g	88 Ra radium 226.0 g	89 Ac actinium 227.0 g	104 Rf Rutherfordium 261.1 g	105 Db dubnium 262.1 g	106 Sg seaborgium 263.1 g	107 Bh bohrium 262.2 g	108 Hs hassium 265 g	109 Mt meitnerium 266 g

58 Ce cerium 140.1 g	59 Pr praseodymium 140.9 g	60 Nd neodymium 144.2 g	61 Pm promethium 144.9 g	62 Sm samarium 150.4 g
90 Th thorium 232.0 g	91 Pa protactinium 231.0 g	92 U uranium 238.0 g	93 Np neptunium 237.0 g	94 Pu plutonium 244.1 g

TABLE

								2 He helium 4.0 g
			5 B boron 10.8 g	6 C carbon 12.0 g	7 N nitrogen 14.0 g	8 O oxygen 16.0 g	9 F fluorine 19.0 g	10 Ne neon 20.2 g
			13 Al aluminum 27.0 g	14 Si silicon 28.1 g	15 P phosphorus 31.0 g	16 S sulfur 32.1 g	17 Cl chlorine 35.5 g	18 Ar argon 39.9 g
28 Ni nickel 58.7 g	29 Cu copper 63.5 g	30 Zn zinc 65.4 g	31 Ga gallium 69.7 g	32 Ge germanium 72.6 g	33 As arsenic 74.9 g	34 Se selenium 79.0 g	35 Br bromine 79.9 g	36 Kr krypton 83.8 g
46 Pd palladium 106.4 g	47 Ag silver 107.9 g	48 Cd cadmium 112.4 g	49 In indium 114.8 g	50 Sn tin 118.7 g	51 Sb antimony 121.8 g	52 Te tellurium 127.6 g	53 I iodine 126.9 g	54 Xe xenon 131.3 g
78 Pt platinum 195.1 g	79 Au gold 197.0 g	80 Hg mercury 200.6 g	81 Tl thallium 204.4 g	82 Pb lead 207.2 g	83 Bi bismuth 209.0 g	84 Po polonium 209.0 g	85 At astatine 210.0 g	86 Rn radon 222.0 g
110 Ds darmstadtium 269g	111 Rg roentgenium 272 g	112 Cn copernicium 277 g	113	114	115	116	117	118

63 Eu europium 152.0 g	64 Gd gadolinium 157.3 g	65 Tb terbium 158.9 g	66 Dy dysprosium 162.5 g	67 Ho holmium 164.9 g	68 Er erbium 167.3 g	69 Tm thulium 168.9 g	70 Yb ytterbium 173.0 g	71 Lu lutetium 175.0 g
95 Am americium 243.1 g	96 Cm curium 247.1 g	97 Bk berkelium 247.1 g	98 Cf californium 251.1 g	99 Es einsteinium 252.1 g	100 Fm fermium 257.1 g	101 Md mendelevium 258.1 g	102 No nobelium 259.1 g	103 Lr lawrencium 262.1 g

Made in the USA
Monee, IL
02 December 2021